Your Body:

4. Muscles and Movement

Dr. Gwynne Vevers

Illustrated by Sarah Pooley

LOTHROP, LEE & SHEPARD BOOKS
New York

OTHER TITLES IN THIS SERIES

Skin and Bone

Blood and Lungs

Feeding and Digestion

First published in Great Britain in 1984 by The Bodley Head
 Inquiries should be addressed to Lothrop, Lee & Shepard Books, a division of William Morrow & Company, Inc., 105 Madison Avenue, New York, New York 10016.
Printed in the United States of America.
First U.S. Edition 1 2 3 4 5 6 7 8 9 10

Library of Congress Cataloging in Publication Data
Vevers, Gwynne, 1916-
Your body.
Includes index.
Contents: 1. Skin and Bone—2. Blood and lungs— [etc.]—4. Muscles and movement.
1. Body, Human—Juvenile literature. 2. Body, Human. I. Pooley, Sarah, ill. II. Title.
QP37.V48 1983 612 83-18757
ISBN 0-688-02825-X
ISBN 0-688-02826-8 (lib. bdg.)

SNORE, SNORE...
Lazy Rob is using his muscles, even when sleeping.

About 40 percent of your body weight is muscle, and there are about 650 muscles of all different shapes and sizes attached to parts of the skeleton. Without muscles you wouldn't be able to make the most simple movements, such as walking, running, or even breathing.

650 muscles — unbelievable!
GOSH!
You weigh 66 pounds (30kg), and so about 26 pounds (12 kg) is the weight of your muscles!
Every fleshy part of your body has muscle. It is more obvious in certain areas —the lower part of the leg, for example.

The muscles in your arms and legs are long and strap-like, with a slight bulge in the middle. They taper at each end and are attached to bone by tough, whitish straps called tendons.

This is a muscle.
This is the tendon. It is where the muscle attaches to the bone. It is very tough because if it was weak, it would just snap... Ping!!
This is a bone.
This is a diagram of the muscles of the arm.

Feel the Achilles tendon behind your ankle. This is the thickest and strongest tendon in the body and is about 6 inches (15 cm) long. It attaches the big muscles of the calf to your heel bone.

The Achilles tendon is named after Achilles. Achilles' mother held him by the heel and ducked him in the river. The legend was that the river would make him safe from harm. His heel, however, wasn't in the water...

The muscles of your head and the main part of your body form broad, overlapping sheets.

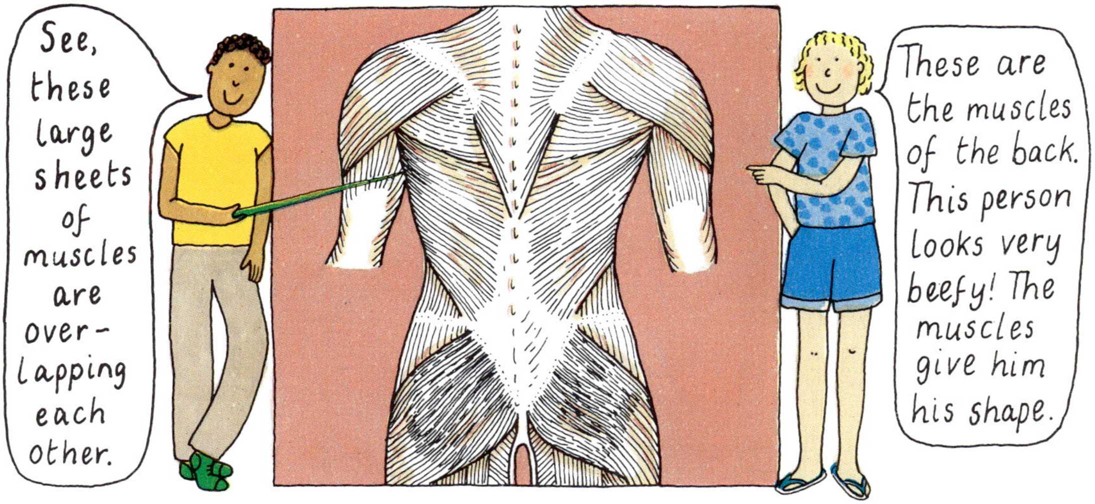
See, these large sheets of muscles are over-lapping each other.
These are the muscles of the back. This person looks very beefy! The muscles give him his shape.

Lie on your back on the floor and raise your feet, keeping your legs straight. Your soft stomach can now be felt as a hard sheet made up of several layers of muscle.

8 inches (20 cm)

Matt is trying this simple exercise. Don't just watch him — YOU try it too. It's EASY!!

There are a great many muscles in your face. Around each eye there is a small circular sheet of muscle which is used, for instance, when you squint your eyes against the sun.

Here is the circular sheet of muscle surrounding the eye.

Mad dogs and Englishwomen go out in the midday sun!

To smile you use about 17 muscles, and to frown, more than 40.

It's certainly a lot more tiring to frown. Brian and I are using 23 more muscles than Chrissie and John.

The hand and fingers have 36 small muscles which allow you to bend and spread your fingers.

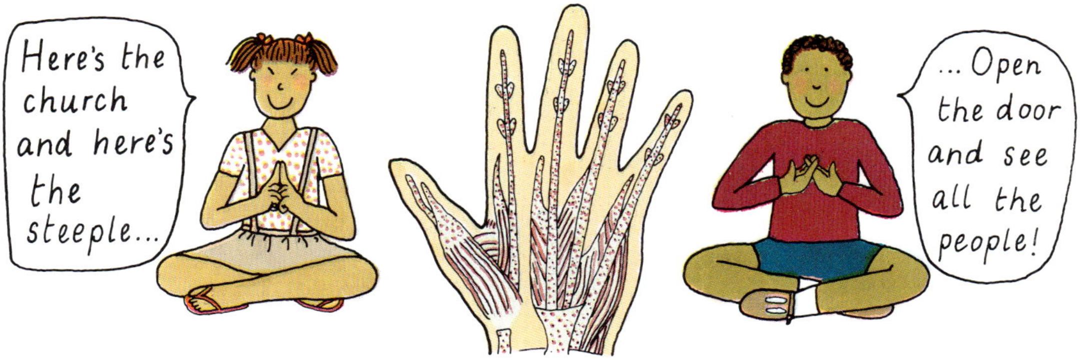
Here's the church and here's the steeple...
...Open the door and see all the people!

These are especially important for pianists, who have to exercise them continually.

I can stretch an octave – that's eight notes. But my little sister's hands can't stretch that far yet!

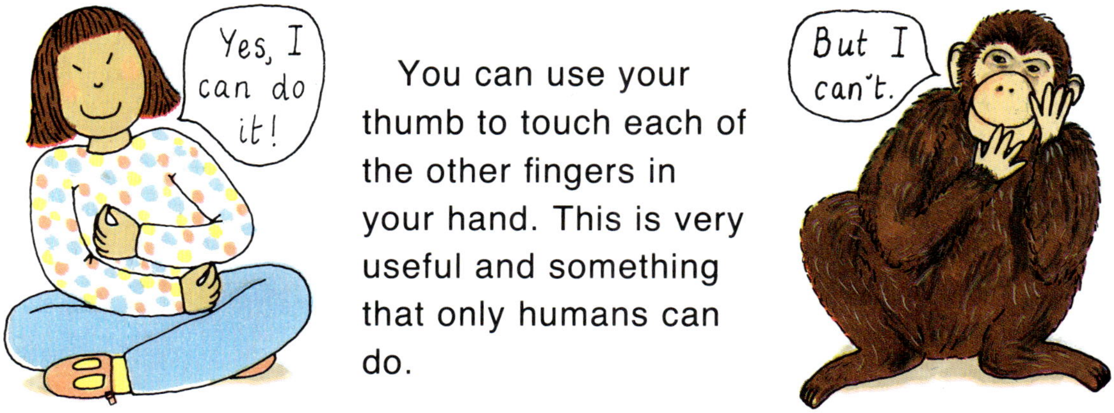
Yes, I can do it!
But I can't.

You can use your thumb to touch each of the other fingers in your hand. This is very useful and something that only humans can do.

Each muscle is made up of thousands of fibers, some only 4 ten-thousandths of an inch (a hundredth of a millimeter) across, some up to 8 inches (20 cm) long. You can see muscle fibers in a piece of steak, where they have usually been cut across, or in a chicken leg.

For my birthday treat, Mom and Dad have taken me to a nice restaurant!
My spécialité is steak au poivre. Look at ze close-up to see ze muscle fibers zat make up ze steak!
CLOSE-UP OF STEAK
muscle fibers

When you move your arm, you are using muscles. The two most important muscles in a human arm are the biceps and the triceps. When you bend your arm, your biceps muscle contracts. At the top of the arm it is attached by a tendon to the scapula, or shoulder blade. At the other end, the biceps is attached to the radius bone in your forearm. When your biceps muscle contracts, it pulls your forearm toward you. This would happen, for example, when you put your hand to your mouth. At the same time, the triceps muscle has relaxed.

This is your biceps contracting or shortening.
The top end of the biceps is attached to the scapula by a tendon.
The arm is being raised.
This is the triceps muscle. When you bend your arm, the triceps relaxes.
The tendon of the biceps attaches to the radius bone in the forearm.

THE HUMAN ARM (RAISED)

HARRY'S
GYM
Left arm straight
Left hand facing upward

You can feel your biceps working. Hold your left arm straight out with the palm of the hand facing upward. Now slowly bend your arm, and with your right hand, feel the left biceps contracting to form a bulge.

HARRY'S
GYM
Arm is slowly bent upward
Right hand feels the bulge made by the left biceps contracting

The triceps muscle starts at the scapula and upper arm bone and is joined at the elbow to the forearm bone. When the triceps contracts, the biceps relaxes and the arm is straightened.

When you are walking, your arms are almost straight and swing along each side. Your arms help you to keep your balance.

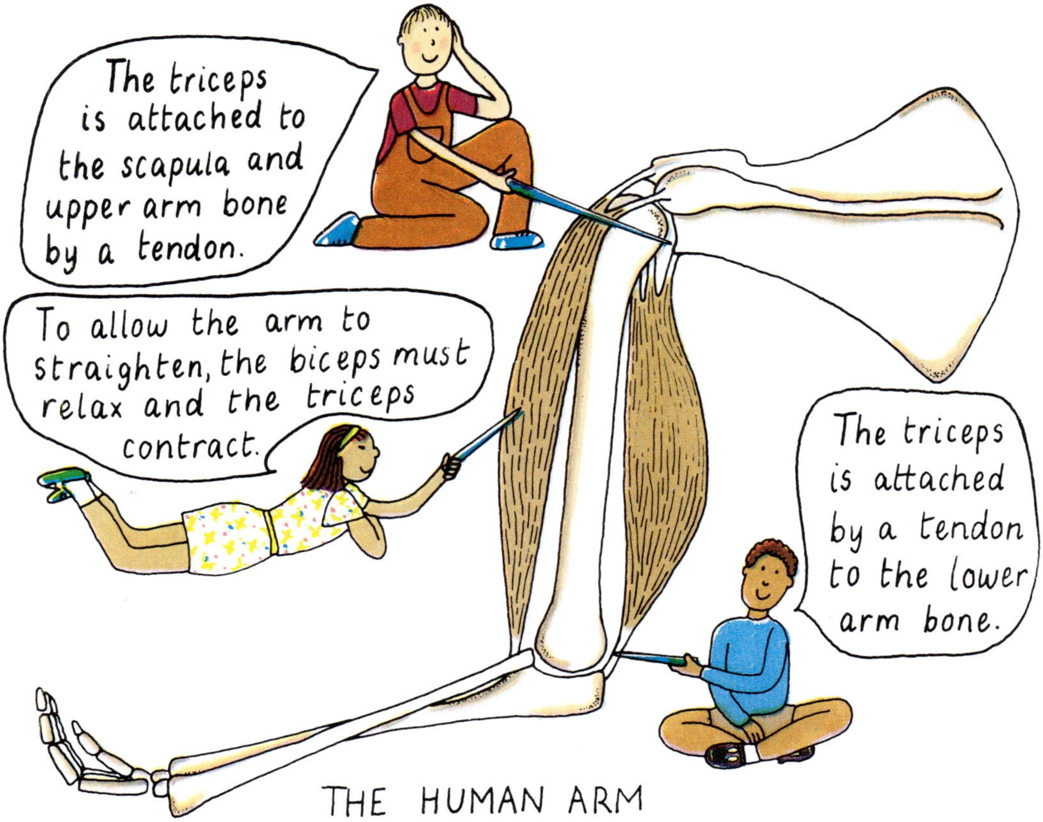
The triceps is attached to the scapula and upper arm bone by a tendon.
To allow the arm to straighten, the biceps must relax and the triceps contract.
The triceps is attached by a tendon to the lower arm bone.

THE HUMAN ARM
(LOWERED)

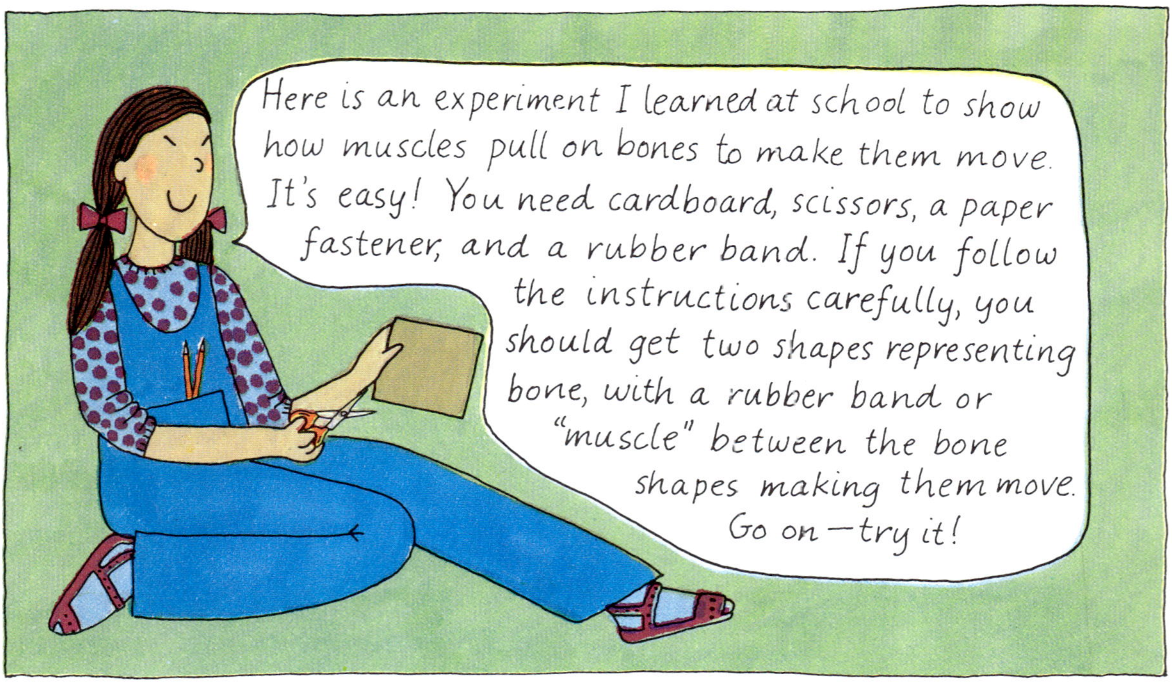
Here is an experiment I learned at school to show how muscles pull on bones to make them move. It's easy! You need cardboard, scissors, a paper fastener, and a rubber band. If you follow the instructions carefully, you should get two shapes representing bone, with a rubber band or "muscle" between the bone shapes making them move. Go on—try it!

EXPERIMENT TO SHOW HOW MUSCLES MAKE BONES MOVE

1

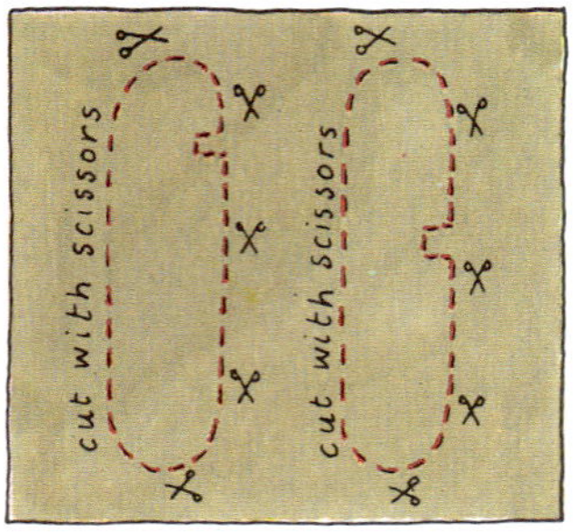
cut with scissors
cut with scissors

Take the cardboard and cut out two identical shapes. Cut a slot in each shape, one near the top and the other in the middle.

2

Take your paper fastener and your rubber band:

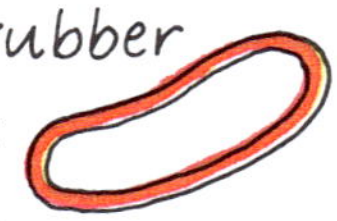

Attach the two shapes together at one end with the paper fastener (you may need to make a hole to make it easier). Attach the band in between the slots.

3 IT SHOULD LOOK LIKE THIS:

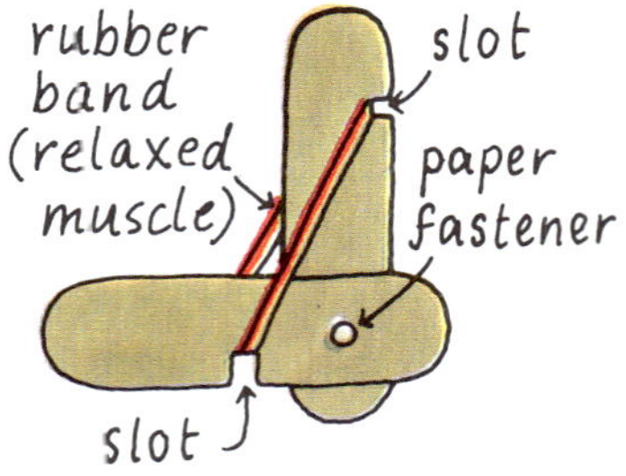
rubber band (relaxed muscle)
slot
paper fastener
slot

4

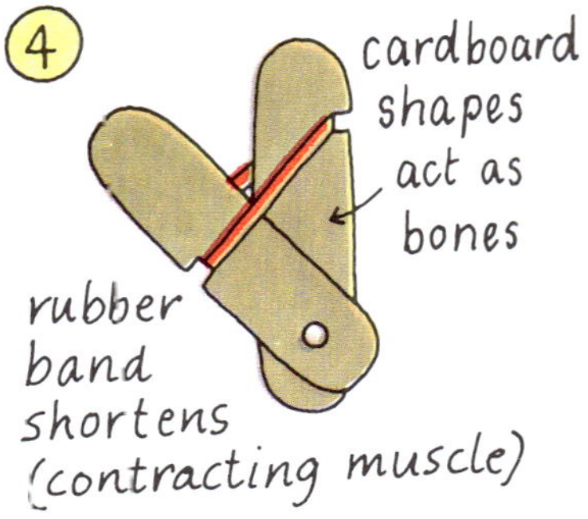
cardboard shapes act as bones
rubber band shortens (contracting muscle)

My uncle Jim is a bricklayer. He lifts bricks all day, so he must be very strong by now! I'd rather stick to lifting lighter things!

Other muscles are involved in the simple action of bending your arm. If you pick up a brick in your hand instead of a peanut, the muscles around the shoulder joint and those of the rib cage may contract to steady your arm. Other muscles in the arm contract to keep your hand steady when it is reaching your mouth.

This is easy!
MS WEIGHTLIFTER '83

Muscles act in groups. When an athlete is getting in shape to run a race, he exercises his muscles in order to get them to act together in the most efficient manner.

Phew! Wheeze... Puff... Puff!
I'm training for the marathon, puff, puff. I've been running every day for a couple of months now. I'm in better shape; my muscles are working very efficiently. Do you think I'll win?!
Hey, Susie! See you at the marathon next week!
THE PARK
Watch for the marathon runners!

On the other hand, when someone has been sitting or lying down for a long time, their muscles may need exercising to make them fully active again.

YAWN!
S-T-R-E-T-C-H
S-T-R-E-T-C-H
I'm raring to go now — my muscles are fully active again

There are three kinds of muscle: skeletal muscle, smooth muscle, and heart muscle.

Skeletal muscles contract when your brain tells them to. When you kick a ball, for example, your brain, in a fraction of a second, has told your leg muscles to work. Skeletal muscles are also known as voluntary muscles because they do what you want them to do.

Brain calling leg muscles, brain calling leg muscles! Kick that ball! Over and out.
RATTLE RATTLE
I'm getting fed up with kicking this ball. If we are only voluntary muscles, can't we go on strike?

INTERNATIONAL SCHOOLS ATHLETICS
Jimmy won his race, but he can't move his legs now! Here, put this sweater on, Jimmy, because otherwise you might cool off too quickly and get chilled.
Look, Jimmy did well, but he looks extremely tired now.
Phew, I'm exhausted. My legs feel dead, but I'll be all right in a minute or two.

Voluntary muscles become tired very quickly. After a race, your legs will feel very tired and more or less useless because the muscles just will not contract any more. They have been doing so much contracting that they have become heated and too tired.

As you sweat, extra heat soon disappears and your body cools off, but put on a sweater so that you don't get too cool too quickly.

Brrr... the sea is very cold today! I must dry myself quickly and put some clothes on.

Even though I've got lots of warm clothes on, I'm still cold and shiver to keep warm.

When you come out of cold water, you shiver without meaning to because your body is losing heat. By making rapid muscle movements you create heat to bring your body temperature back up to normal. You can also shiver on purpose, when you try to keep warm on a cold day.

Ooooh! I've got cramps. Quick, save me, someone!
Sometimes our muscles get painful spasms called cramps. Jack has got them in the swimming pool, but Gina will save him.
Sometimes we get cramps in our legs in bed or anywhere!

Smooth muscles are found in various internal organs of the body, especially in the walls of the bladder, of the blood vessels, and of the alimentary or food canal. When food has been swallowed, it is pushed along by repeated muscle contractions in the canal wall. This is called peristalsis.

This is a lump of food.
I can feel the lump of food move as I swallow.
This is the muscular wall of the food canal. See how it contracts at intervals to push the food along!

These contractions are not controlled by willpower. So smooth muscles are also known as involuntary muscles.

The thick, powerful muscle that makes up the walls of the heart chambers is called heart muscle. It goes on contracting and relaxing all the time, pumping blood around the body throughout your whole life. Heart muscle normally contracts and relaxes about 70 times a minute.

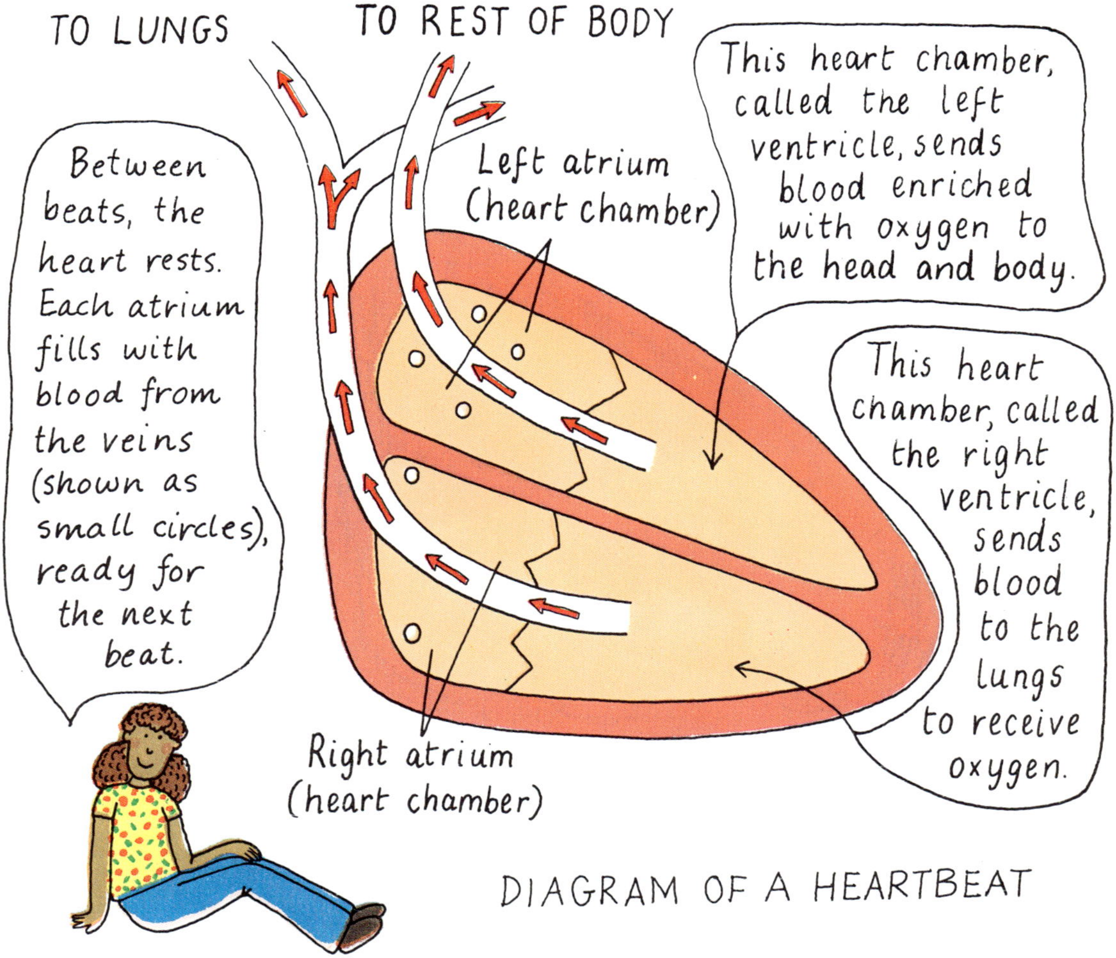

TO LUNGS
TO REST OF BODY
This heart chamber, called the left ventricle, sends blood enriched with oxygen to the head and body.
Left atrium (heart chamber)
Between beats, the heart rests. Each atrium fills with blood from the veins (shown as small circles), ready for the next beat.
This heart chamber, called the right ventricle, sends blood to the lungs to receive oxygen.
Right atrium (heart chamber)

DIAGRAM OF A HEARTBEAT

Muscles of all kinds are supplied with blood vessels and with nerves that come from the brain and the spinal cord. The blood provides the muscles with energy, and the nerves tell the muscles when to contract.

THE SPINAL CORD AND SPINAL NERVES

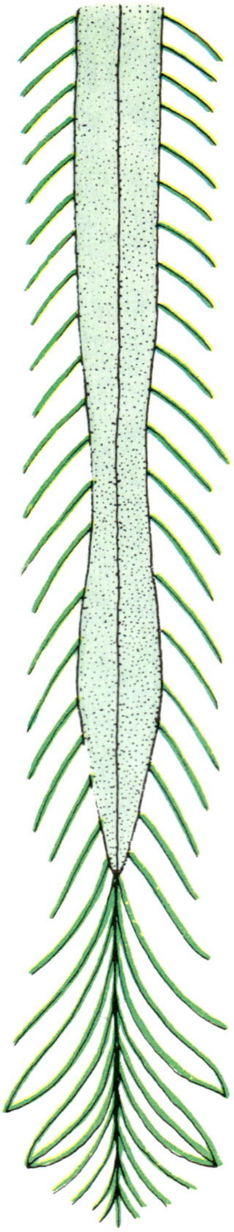

The spinal cord is about 18 inches (45 cm) long and as thick as your little finger!

THE BRAIN SEEN FROM BELOW

There are 12 pairs of brain nerves.
spinal cord

When I cross the road, I look left and right and left again.
And I listen very hard, just in case!
When it is safe to cross, the messages from my eyes and ears tell my brain. My brain then instructs my nerves to tell the muscles to contract.
Guess what! I then walk across the road. Isn't it a complicated thing to do!?!

Your brain receives messages from your eyes and ears and uses the nerves as a telephone system to tell your voluntary muscles when to contract.

Mom's a telephone operator, and she explained that your brain is like a telephone system. The brain sends messages to the nerves, which then tell your muscles to contract. Mom plugs in the wires, and people get their messages.

A brain injury can be very serious because the vital nerves do not heal very quickly, and so messages cannot reach muscles in other parts of the body, which may become paralyzed.

Charlie and I are both paralyzed in our legs. When I was little, I got polio, which is a disease that damages the nerves. Charlie was involved in a motorbike accident. He hit his head and damaged his brain.
Just because we are in wheelchairs, it doesn't mean we don't have fun! Yvonne and I are going to have a race. On your mark, Yvonne... Ready, Set, GO!!

Do you like our little show?
Look! No hands.
Watch us balance!
I'm Olga Korbut the second!
This is great!

With the nerves and muscles working together, your body is able to jump and swim, turn somersaults, and ride a bicycle. Even sitting on your bottom uses muscles.

Index

Page numbers in *italic type* indicate illustrations.

Achilles tendon, *4,* 4
Arms, *4,* 4, *9,* 9-13, *11*

Biceps, *9-11,* 9-11
Blood vessels, 18, 20
Body temperature, *17,* 17
Bones, *4,* 4, *9,* 9-13, *11*
and muscle, experiment, *12*
Brain, 15, *20-22,* 20-22

Cramps, *17*

Exercise, *14,* 14
Eyes, *6,* 6

Face, 6
Fingers, *7,* 7
Food canal, *18,* 18
Frowning, *6,* 6

Hands, *7,* 7, 10
Heart, *19,* 19
Heartbeat, *19*
Heart muscle, 15, 19

Involuntary muscle, 18

Legs, 4, *15-16,* 15-16

Muscle contraction, *9-12,* 9-13, 15-16, 18-21
Muscle fibers, *8,* 8
Muscles:
of arms, *4,* 4, *9,* 9-13, 11
and body weight, *3,* 3
and bone, experiment, *12*
group action of, 13-14
layers of, *5,* 5
number of, *3,* 3
types of, 15-19
See also specific body part

Nerves, *20,* 20-23, *21*

Paralysis, *22,* 22
Peristalsis, 18
Polio, *22*

Radius, *9,* 9

Scapula, *9,* 9, *11,* 11
Shivering, 17
Skeletal muscle, 15
Smiling, *6,* 6
Smooth muscle, 15, 18
Spinal cord, *20,* 20
Stomach, 5
Sweating, 16

Tendons, *4,* 4, *9,* *11*
Thumb, *7,* 7
Triceps, *9,* 9, *11,* 11

Voluntary muscle, *15,* 15-16, 21

Walking, *11*